Satellites

Rebecca L. Johnson

LERNER PUBLICATIONS COMPANY
MINNEAPOLIS

For Dano, who used to stand with me in the
backyard looking up at the stars,
and who still searches the night sky

Lerner Publications Company
A division of Lerner Publishing Group, Inc.
241 First Avenue North
Minneapolis, Minnesota 55401 U.S.A.

Website address: www.lernerbooks.com

Library of Congress Cataloging-in-Publication Data

Johnson, Rebecca L.
        Satellites / by Rebecca Johnson.
            p.   cm. — (Cool science)
        Includes bibliographical references and index.
        ISBN–13: 978-0-8225-2908-8 (lib. bdg. : alk. paper)
        ISBN–10: 0-8225-2908-4 (lib. bdg. : alk. paper)
        1. Artificial satellites—Juvenile literature. I. Title. II. Series.
        TL796.3.J64 2006
        629.46—dc22   2004030298

Manufactured in the United States of America
2 – JR – 10/1/09

Table of Contents

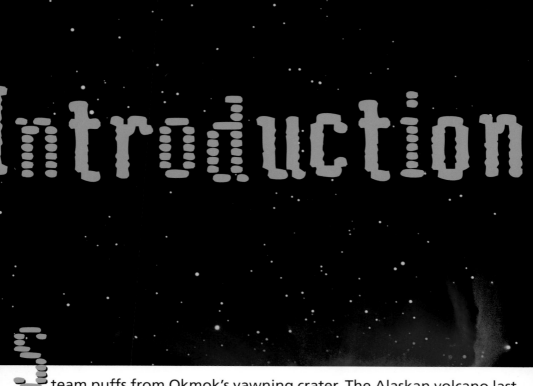

Introduction

Steam puffs from Okmok's yawning crater. The Alaskan volcano last erupted in 1997. When will it erupt again?

In a helicopter that's circling the crater, a scientist is trying to answer that question. He's studying satellite images of the volcano. The pictures show in rainbow colors that the ground around Okmok is beginning to bulge. Magma (melted rock beneath Earth's surface) is on the move, pushing up from below. You'd never notice these bulges at ground level. They're just a few inches high. But satellites can detect this slight change. And it's a sign that Okmok may erupt again soon.

Far out in the Pacific Ocean, a ship sails across blue tropical waters. There's no land in sight. But the captain knows exactly where the ship is, thanks to a special navigation system. The instrument has just locked on to five satellites orbiting Earth. It shows the ship's precise location—accurate to within a foot or two (less than a meter).

This satellite image shows Okmok volcano after its 1997 eruption.

In his office, an astronomer examines the latest pictures from the Hubble Space Telescope. One is a dazzling image of a vast cloud of dust and gas. It's all that's left of a massive star the exploded millions of years ago, millions of light years from Earth. What a sight! It's a view of our universe that only a satellite can provide.

Three different people are doing different things in different places. What do they have in common? Satellites play a big role in their lives!

FUN FACT!

Did you know that there are close to 5,000 artificial satellites flying around in space today? About half of those are functioning (satellites don't last forever!). Most belong to the United States and Russia. But 26 other countries and organizations, including France, China, Japan, India, and the European Space Agency, can lay claim to the rest.

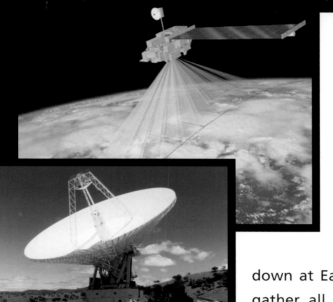

Right now, several thousand satellites are circling Earth high overhead. If you're lucky and have sharp eyes, you might spot a few among the stars on a clear night. But what you can see is nothing compared to what the "eyes" of a satellite can see. Some look down at Earth. They take pictures and gather all sorts of information about our planet. Others look out into space, toward the planets, the Sun, and the unknown worlds beyond our galaxy.

A satellite helps transmit voice and image signals (*top*). The signals are received by antennas (dishes) on Earth (*middle*). A space satellite took this image of a supernova, or exploding star (*bottom*).

Satellites can do more than just "see." They also make it possible for you to phone a friend in another country, watch your favorite television shows, and find out what tomorrow's weather will be like. They can even help you find your way back to camp if you get lost in the woods.

Satellites touch your life in more ways than you may realize. And they do it all from space.

chapter 1

What's a Satellite?

So, just what *is* a satellite? Technically speaking, a satellite is any smaller object that follows an orbit around a bigger object. An orbit is a regular, repeated path that one object takes around another object in space. Some satellites are natural. The Moon is a satellite of Earth. And Earth is a satellite of the Sun, as are the other planets in our solar system.

But what we're talking about here are artificial satellites. These are human-made instruments that are launched into space, where they orbit Earth or other objects.

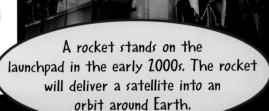

A rocket stands on the launchpad in the early 2000s. The rocket will deliver a satellite into an orbit around Earth.

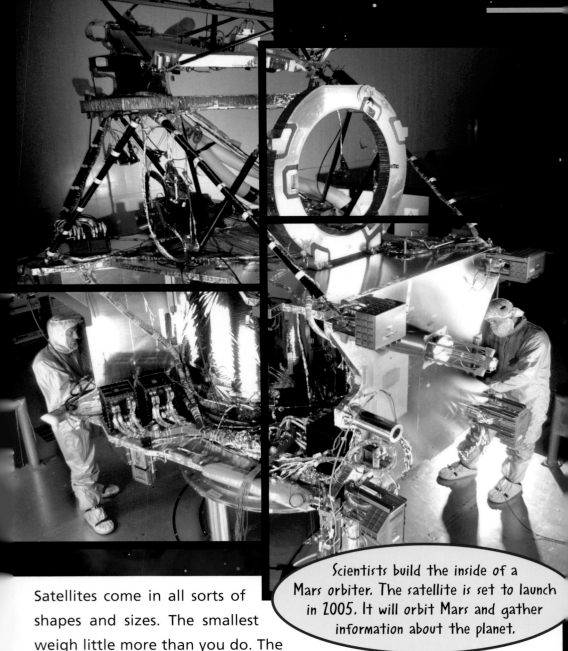

Scientists build the inside of a Mars orbiter. The satellite is set to launch in 2005. It will orbit Mars and gather information about the planet.

Satellites come in all sorts of shapes and sizes. The smallest weigh little more than you do. The largest tip the scales at thousands of pounds.

Satellites differ in other ways too. Some simply reflect radio signals that are beamed up at them. Others are packed with sensors, instruments, and computers that gather huge amounts of information and then transmit it down to Earth.

Getting There

Big or small, getting a satellite into orbit is no easy task. To do that, you need a launch vehicle. That's usually a rocket with powerful engines that burn chemical fuel. A rocket blasts off from Earth with enough speed and power to escape Earth's atmosphere and enter the vacuum of space. How much is enough? The rocket needs to reach an altitude (height) of at least 120 miles (200 kilometers) and a speed of more than 18,000 miles per hour (29,000 km/hr)!

Hitchin' a Ride

Small satellites can be launched from airplanes that fly at very high altitudes. Another, smaller launch vehicle is still needed to carry the satellite out into space. Sometimes satellites can hitch a ride into space on a space shuttle. The onboard astronauts launch the satellites themselves. They can also pick up broken satellites and try to fix them in space or bring them back to Earth for repairs.

An astronaut prepares a satellite for launch from the cargo area of the space shuttle (below). Astronauts attach satellites to a robotic arm (right), which they then use to place satellites in space.

All satellites follow some type of orbit. Orbits are the result of a perfect balance between two forces: motion and gravity. As a satellite moves through space, it has forward motion. The object the satellite is orbiting has gravity. Gravity is a pulling force that tugs at the satellite. If a satellite's forward motion is stronger than the gravity pull from the object, the satellite will zoom off into space. If the gravity pull is stronger than the satellite's motion, the satellite will crash into the bigger object. As long as motion and gravity are in perfect balance, however, an orbit is maintained.

In its orbit of Earth, a satellite passes over the island of Crete and the Mediterranean Sea. Satellites follow different kinds of orbits.

Satellites follow different orbits, depending on what they are designed to do. Satellites in polar orbits circle Earth from pole to pole. Satellites that survey most of Earth as it turns below them have polar orbits. Many weather satellites and those that survey land areas follow polar orbits.

A satellite in a geostationary orbit lies directly above the equator. It stays at an altitude of 22,300 miles (35,888 km). At that height, a satellite completes one orbit of Earth in the same amount of time that it takes Earth to rotate once. As a result, satellites in geostationary orbits stay above the same point on Earth's equator all the time.

Some satellites follow an elliptical, or oval, orbit rather than a round one. Elliptical orbits are best for satellites that take scientific measurements at different altitudes above Earth's surface.

Orbits: What Comes Around, Goes Around

Circular, polar, and elliptical orbits can be low (relatively close to Earth's surface) or high (farther out in space). In order to be completely free of Earth's atmosphere, a satellite needs to orbit at least 180 miles (300 km) above sea level. In orbits lower than that, the satellite bumps into air molecules. These collisions, although tiny, slow the satellite down. If a satellite loses too much speed, the pull of Earth's gravity will overcome its forward motion. The satellite will fall out of orbit and crash into the planet's surface.

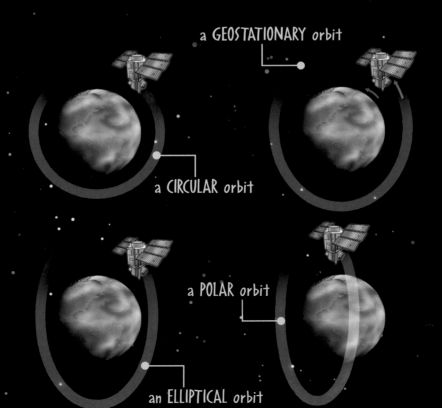

a GEOSTATIONARY orbit

a CIRCULAR orbit

a POLAR orbit

an ELLIPTICAL orbit

Once in orbit, it's important for satellites to stay there. Natural forces in space, such as the pull of Earth's gravity, may cause a satellite to veer off course. Many satellites are equipped with tiny rockets called thrusters. By controlling which thrusters fire and for how long, engineers on Earth can keep satellites in proper orbit. They can also turn satellites so that onboard instruments are pointing in the right direction.

Solar cells provide the power to run a satellite's cameras and other instruments. Solar cells gather energy from sunlight and convert it into electricity. Hundreds of solar cells are arranged to form large solar panels. Solar panels usually have a battery backup for times when sunlight can't reach a satellite.

Blinded by Science: Solar Power

Solar panels on a satellite can be pretty big, especially on satellites that use a lot of electricity. The solar panels on the *Hubble Space Telescope*, for example, cover

an area of about 3,120 square feet (290 sq. m). A satellite's solar panels are typically folded up during its journey into space. When the satellite reaches orbit, the panels unfold like a butterfly emerging from its cocoon and spreading its wings.

A technician tests the circuitry (wiring) of a satellite's solar panel. The massive solar panel will power the satellite while it is in orbit around Earth.

An astronaut rides the robotic arm of a space shuttle to repair a satellite over Earth.

Surviving in Space

Many satellites look surprisingly fragile. But to survive in space, they need to be pretty tough customers. A satellite is exposed to intense heat when it's facing the Sun. To keep from burning up, many satellites have special panels that reflect heat into space. Some spin as they travel so that the searing rays of the Sun don't strike one spot for more than a few seconds.

Satellites also get hit by sand-grain-sized meteors. These tiny flying objects, called micrometeoroids, can tear through metal. Micrometeoroids can damage a satellite's body as well as its solar panels and instruments. In order to last, satellites must be built to survive micrometeoroid hits.

With all the dangers that satellites face out in space, it's amazing that so many of them keep working for years—even decades—without fail. The first satellites didn't last nearly as long as modern-day satellites do. Simply getting a satellite into space was the big challenge. But when it finally happened, the world was forever changed.

chapter 2

The Race into Space

Humans have long thought about artificial satellites. But the idea of launching satellites didn't really take off until 1952. That's the year a group of scientists challenged the world to get a satellite into space. You can imagine the effect that such a global challenge had. Countries saw it as a competition. And the two biggest competitors at the time were the United States and the Union of Soviet Socialist Republics—the USSR (the Soviet Union, or Russia).

As it turned out, the Soviets beat the Americans—and everyone else—into space. On October 4, 1957, the Soviets launched the world's first artificial satellite, called *Sputnik I.* By today's standards, it was a pretty simple satellite. *Sputnik I* was little more than a thermometer, a battery, and a radio inside of an aluminum sphere slightly larger than a basketball. The whole satellite weighed just 183 pounds (83 kilograms). As *Sputnik I* orbited Earth in an elliptical path, its radio sent a steady, beeping signal to receiving stations on the ground.

A Soviet scientist works on *Sputnik I*. The Russians shocked the world with their launch of the satellite on October 4, 1957. *Sputnik I* circled Earth in about an hour and a half in an elliptical orbit.

A New Age

Sputnik I's launch caught the world by surprise. The U.S. space program, then run by the military, had been working on its own satellite. But the project was behind schedule. The sudden appearance of *Sputnik I* was embarrassing for U.S. space scientists.

For many Americans, however, the fact that the Soviets had a satellite orbiting Earth was more than embarrassing. It was terrifying. People feared that if the Soviet Union could launch a satellite, they could also launch nuclear weapons.

The launch of little *Sputnik I* was a turning point in history. It marked the start of the Space Age. And it started a space race between the United States and the Soviet Union that would last for many years.

Sputnik Strikes Again

Space scientists in the United States worked as fast as they could to get the first U.S. satellite into orbit around Earth. But before they succeeded, the Soviet Union struck again. The Russians launched *Sputnik II* on November 3, 1957.

Sputnik II was much larger than *Sputnik I.* The new satellite weighed more than 1,000 pounds (450 kg). But far more remarkable was its cargo. *Sputnik II* was the first satellite to carry a living creature into space: a dog named Laika.

Soon after launching *Sputnik I*— the first human-made satellite in space—the Soviets launched *Sputnik II (above).* The Satellite carried a dog—the first living creature in space.

Earth's Magnetic Field

Like a giant bar magnet, Earth has two magnetic poles. Lines of magnetic force run between these poles. They form what's called Earth's magnetic field, or magnetosphere. Charged particles (negatively charged electrons and positively charged ions) moving through space behave in interesting ways when they hit the magnetosphere. Some can remain trapped in Earth's magnetic field, traveling around and around in what are called radiation belts. Long before *Sputnik I* was launched, scientists thought that electrons and ions could be trapped by the planet's magnetosphere. But they had no proof that any trapped particles were actually there. An instrument aboard a U.S. satellite detected these particles, giving scientists the proof they needed.

Gravity: What Goes Up, Must Come Down

Back in the 1600s, English scientist Sir Isaac Newton (below, left) became the first person to describe the laws of gravity and motion. He also was first to suggest that artificial satellites might one day orbit Earth. In 1903, Russian scientist Konstantin Tsiolkovsky showed mathematically how a satellite could achieve an orbit around the planet. In 1948, Soviet scientist Mikhail Tikhonravov proposed actually making a satellite. Most Soviet officials were skeptical. But a few supported Tikhonravov. The result was *Sputnik I*.

Explorer Takes Flight

The *Sputnik II* launch pushed the United States to work harder and faster. On January 31, 1958, the United States launched the first successful American satellite, *Explorer I*.

Explorer I carried something the first two Soviet satellites had not: a scientific instrument. On board was a small device designed to measure radiation around Earth. It had been built by James Van Allen, a physicist from the University of Iowa. Data (information) sent to Earth by this small radiation detector proved certain characteristics of Earth's magnetic field. The data also led to the discovery of what came to be called the Van Allen radiation belts that encircle our planet.

FUN FACT!

The Sputnik satellites led the United States to create the National Aeronautics and Space Administration, or NASA, in 1958. NASA's first task was to develop a plan for human space exploration. The agency has continued to explore space on many fronts ever since.

Beyond Explorer

After *the Sputnik satellites* and *Explorer I,* both the United States and the Soviet Union—and later other countries—continued to put satellites into orbit. NASA focused its efforts on getting a bird's-eye view of Earth from space. It also focused on studying something that affects everyone every day: the weather.

In the spring of 1960, NASA's Television and Infrared Observation Satellite *(TIROS-1)* sent back the first pictures of Earth as seen from space. People saw their world as never before. They could see entire continents and oceans. They could see cloud formations and storms developing half a world away. Up until that moment, our view of the world had been limited to whatever chunk we could see of it from the ground or from airplanes. But thanks to *TIROS-1,* our worldview suddenly expanded to include the whole planet, spinning slowly through space.

TIROS-1 stayed in orbit for only 78 days. During that time, though, the satellite captured more than 22,000 pictures of Earth. Meteorologists (scientists who study the weather) started using those images immediately to improve weather forecasts worldwide.

Scientists work on *TIROS–I.* This early U.S. satellite gave the world the first pictures of Earth.

Satellite for Sale—Cheap

Satellites don't last forever. Some stop working when their instruments fail. Others fall out of orbit. A satellite that falls out of orbit (and can't be put back) will spiral closer and closer to Earth. It will eventually burn up as it reenters the atmosphere. Broken satellites that stay in space are considered "space junk."

Two astronauts bring back a broken satellite to a space shuttle. Adding a sense of humor to their mission, the astronauts hold a "For Sale" sign in front of the "space junk."

Phone Calls from Space

In the early 1960s, U.S. president John F. Kennedy set the goal of sending an American to the Moon by the end of the decade. He also wanted the United States to develop the first satellite communications system. The idea behind a communications satellite was to send a signal into space. Once in space, the signal could bounce off a satellite and head back down to another spot on the globe. This would make it possible to communicate quickly over long distances. If it worked, people would not have to be connected by telephone or telegraph lines.

NASA had already launched *Echo,* the country's first communications satellite. *Echo* was a giant metal balloon that simply bounced signals back to Earth. The technology was still pretty simple. But for the first time in history, a spoken voice traveled into space and was then sent back to Earth by satellite.

Telstar I (above) was one of several communications satellites planned by the AT&T phone company. Only two made it into space. Telstar I was launched on July 10, 1962. Telstar II went up on May 7, 1963.

The next big step came in 1962, with the launching of the next generation of communications satellite: *Telstar I. Telstar I* didn't just bounce a signal back to Earth. It received the signal, amplified it (increased its strength), and sent it back—more clearly and over greater distances than *Echo* could. To mark the occasion, the chairman of the phone company AT&T used *Telstar* to place a two-minute phone call to Vice President Lyndon Johnson.

Telstar I and Early Bird: London Calling

Telstar I had one glitch. It wasn't in a geostationary orbit. That meant that each time the satellite disappeared below the horizon, its signal was lost until it reappeared overhead. *Early Bird,* the world's first commercial communications satellite, solved this problem. Launched by the International Telecommunications Satellite Organization (Intelsat) in 1965, *Early Bird* was in a geostationary orbit. It transmitted signals for telephone, television, and telegraph across the Atlantic Ocean without interruption.

The test pattern of a British network received in the United States via Telstar I (below) shows that the television signal and the satellite are working. But Telstar I's orbit caused it to drop connections. Scientists corrected this orbit problem in Early Bird (below, right).

Carried on radio waves, the chairman's voice traveled 3,500 miles (5,600 km) through space as it was picked up by *Telstar I* and sent back to Earth. That same day, *Telstar I* carried the first live television transmission between Europe and the United States.

The *Nimbus A (below)* was the first advanced weather satellite. Built from more than 40,000 parts, *Nimbus A* collected information about Earth's weather patterns. The satellite was launched August 28, 1964.

The Nimbus Program

In 1964, NASA launched the first of many *Nimbus* satellites. These satellites were packed with scientific instruments for studying Earth's oceans and atmosphere. Scientists wanted to study the ozone layer that protects Earth from the Sun's deadliest rays.

Going with GOES

The first Geostationary Operational Environmental Satellite (GOES) was launched in 1975. It was the first of many GOES satellites designed to work together around Earth to keep an eye on the world's weather 24 hours a day, seven days a week.

One of the instruments aboard *GOES I* provided day and night images of cloud conditions around the globe. The satellite could continuously monitor dangerous weather events such as hurricanes. It could also relay weather data from more than 10,000 different locations on Earth's surface to a central processing center. There, weather data were plugged into computer models to create weather forecasts that were more accurate than ever before.

A New Look at the Land

In the early 1970s, more satellites appeared that were designed to study Earth in new ways. The first *Landsat* satellite was launched in 1972. As its name implies, *Landsat 1*'s focus was to survey Earth's landmasses from space, photographing everything from tiny islands to entire continents.

Landsat 1's images revealed parts of Earth's surface that no one had ever seen before. The satellite sent back detailed images of remote mountain ranges, forests, glaciers, and deserts. Scientists also got some of their first satellite views of coastlines, vast areas of cropland, and sprawling cities. By comparing *Landsat 1* images from year to year, scientists could track changes to Earth's surface as never before.

Landsat: I Can See Your House from Here

Images captured by Landsat satellites and similar satellites aren't simply photographs like those made with a camera. The sensors aboard these satellites can "see" things that our eyes cannot, such as infrared radiation (heat energy) and microwaves. Images created by sensing infrared radiation can reveal the temperature of lakes and oceans and show whether or not a forest or field is healthy. Microwave sensors can see through clouds. Scientists often use computers to improve the images created by these types of sensors.

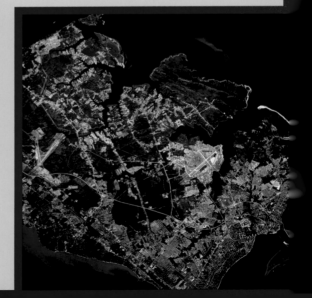

An early Landsat satellite took this photograph of the state of Virginia in April 1973.

Oceans cover three-quarters of Earth's surface. *Landsat* satellites were helping scientists study the land from space. *Seasat 1* became the first satellite specifically designed to study what was going on in the oceans of the world.

Launched in 1978, *Seasat 1* collected data about ocean winds and water temperatures. *Seasat 1* recorded wave heights and lengths and kept an eye on the sea ice in polar regions. *Seasat 1* functioned for only 116 days in space. But in that time, it showed how effective satellites could be in monitoring ocean conditions over large areas. That was impossible to do using ships or even airplanes.

An artist drew this picture of *Seasat 1* in the late 1970s *(right)*. *Seasat 1* made this map of the South Pacific Ocean *(below)*. Scientists added the location labels.

AUSTRALIA

NEW CALEDONIA BASIN

LORD HOWE RISE

BRIDES TR

RIDGE

CHALLENGER PLATEAU

N. I.

TONGA

TONGA

TASMANIA

NEW ZEALAND

S. I.

CHATHAM RISE

BOUNTY CHANNEL

LOUISVILLE RIDGE

GEORGE V FZ

TASMAN FZ

BALLENY FZ

MACQUARRIE RIDGE

CAMPBELL PLATEAU

BOLLONS TABLEMOUNT

UDINTSEV FZ

Totally Radarsat

Radarsat 1 was the first satellite launched by two countries working together, the United States and Canada. Launched November 4, 1995, it was equipped with a microwave sensor that made it possible for the satellite to see through clouds and in darkness. These micro-wave sensors capture very detailed pictures of some of the most remote places on Earth. Radarsat 1 gave the world its first complete look at the vast, frozen continent of Antarctica.

This Radarsat 1 image shows the Weddell Sea near the Antarctic Peninsula. Launched in 1995, the Canadian satellite has the ability to "see" through clouds.

TIROS-1, Telstar I, Nimbus A, Seasat 1, and other early satellites gave us entirely new perspectives on our world. They paved the way for the many satellites that would follow. These satellites are constantly making new discoveries about Earth, our solar system, and the universe.

chapter 3

Up There, Looking Down

In less than 50 years, satellites have opened up entirely new fields, even worlds, to scientists. Satellites have gathered vast numbers of images and huge amounts of information about Earth, other planets, the Sun, our galaxy, and other galaxies far out in space. Satellites are even helping us answer questions such as how the universe began and when and how it may come to an end.

New satellites are being launched all the time. Many of these satellites are designed to help scientists learn more about our planet and how it works. What's up there right now looking down at Earth? Let's take a quick "trip" into space to find out.

A weather satellite took this photograph of Earth in 1992. In the image, Hurricane Andrew spins toward the Louisiana coast.

And Around and Around They Go

NASA worked hard to meet its goal of sending people into space aboard a variety of satellites. In 1962, John Glenn became the first U.S. astronaut to orbit Earth. Glenn orbited the globe three times in a spacecraft capsule nicknamed *Friendship 7*. The mission lasted five hours. By 1969, space technology had advanced far enough to land the first humans on the Moon. Since 2000, astronauts from many countries have lived and worked together aboard one of the largest and most complex satellites ever built: the International Space Station.

John Glenn enters the space capsule *Friendship 7* in 1962. Glenn was the first U.S. citizen to orbit Earth.

Communications Satellites

Most of us take it for granted that we can communicate with almost anyone from any place in the world. Trekking across the wilds of Africa or the ice-covered plains of Greenland? Pull out a satellite phone, and you can chat with a friend in a flash. You can even send e-mail from your laptop!

The Intelsat network of satellites helps make this instant link between people in far-flung parts of the world. *Intelsat* is short for International Telecommunications Satellite Organization. This is an international group that operates a network of 19 communications satellites. Every day, Intelsat satellites transmit billions of communications to 600 ground stations in roughly 150 different countries.

Weather Satellites

In the United States, the National Oceanic and Atmospheric Administration (NOAA) operates two main types of weather satellites. Two GOES satellites are currently orbiting Earth. They are in geostationary orbits over the equator. One of the satellites monitors North and South America and most of the Atlantic Ocean. The other keeps an eye on part of North America and the Pacific Ocean.

Using data from GOES satellites, meteorologists can spot and track tornadoes, flash floods, hailstorms, and hurricanes. Scientists also use GOES images to monitor dust storms, volcano eruptions, and forest fires.

Along with the GOES satellites, two polar-orbiting satellites constantly circle Earth from pole to pole. These are the Advanced Television Infrared Observation Satellites. They are descendants of the original TIROS satellites. Sensors aboard these satellites collect information about the atmosphere, cloud cover, and sunlight.

A GOES satellite (*above*) sends weather information to meteorologists on Earth in the early 2000s. A GOES satellite captured this image of deadly Hurricane Fran (*below*) off the U.S. East Coast during the 1990s.

Satellites to the Rescue

In addition to keeping an eye on the weather, polar-orbiting weather satellites are part of a worldwide satellite rescue network. These satel-

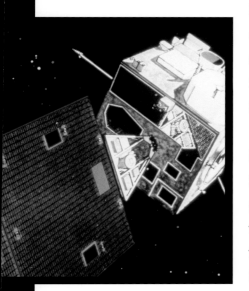

lites help rescuers locate people who are lost and in need of help on land or at sea. The satellite network is called COSPAS-SARSAT. *SARSAT* stands for Search and Rescue Satellite Aided Tracking.

More than 30 nations belong to the COSPAS-SARSAT network. They operate dozens of ground stations and mission control centers that can quickly organize search and rescue teams. Imagine you were on a boat with a broken motor, drifting in the middle of the ocean hundreds of miles from land. COSPAS-SARSAT could find you and send help. The key to making all this work is the satellites.

SARSAT: Don't Go to Sea without It

SARSAT had its beginnings in 1970. That year an airplane carrying two members of the U.S. Congress crashed in a remote region of Alaska. All attempts to find the downed plane failed. After this tragedy, Congress passed a law that all aircraft in the United States carry an emergency locator device. The device sends out a signal to help rescuers find hurt or stranded people in similar situations. Over time, satellites became the tools that could best pick up these distress signals and transmit their location to rescue centers.

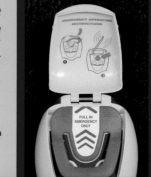

A SARSAT beacon (right) sends a signal to SARSAT satellites showing the location of the user.

COSPAS-SARSAT System Overview

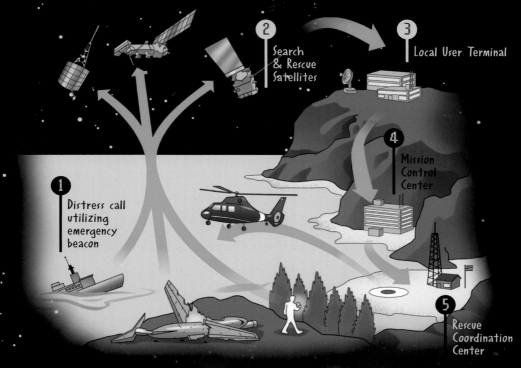

② Search & Rescue Satellites

③ Local User Terminal

④ Mission Control Center

① Distress call utilizing emergency beacon

⑤ Rescue Coordination Center

Rescue satellites include several NOAA weather satellites and satellites operated by Russia, India, and the European Union. Instruments aboard these satellites pick up distress signals and transmit information about exactly where the signals are coming from. Since the COSPAS-SARSAT network was formed in 1982, more than 17,000 people have been rescued worldwide. Nearly 5,000 of those were in the United States.

Navigation Satellites

Have you heard of GPS? The letters stand for Global Positioning System. GPS uses a network of at least 24 orbiting satellites to pinpoint locations on Earth. The system is used on airplanes to navigate through the air, on ships to cross oceans, and in cars to navigate city streets. Geographers and geologists use GPS to find and mark locations on the landscape. And ordinary people who like to hike, camp, and climb use it to find their way as they tramp through wilderness areas.

How does the GPS work? Imagine you're walking in the woods and you're not quite sure how to get back to your campsite. Luckily for you, you're carrying a GPS receiver, about the size of a cell phone. And luckily, you remembered to enter the position of your camp into the GPS unit before you set out on your hike.

When you turn on your GPS unit out in the woods, it locks on to several GPS satellites. GPS satellites circle Earth, emitting continuous navigation signals. Your GPS unit picks up signals from three to five of these satellites that are passing overhead. It uses the signals to calculate your location on Earth. It is accurate to within a few feet. On its little screen, your GPS unit shows you exactly where you are so you can figure out which direction to go to get back to camp.

GPS: Wherever You Go, There You Are

The U.S. military developed the Global Positioning System. It is managed by the U.S. Air Force. The military allowed civilians to start using the GPS in 1993. However, the signals that can be picked up by civilian (nonmilitary) GPS receivers are not quite as accurate as those used by the military. The European Space Agency is currently developing another kind of GPS that will be based on a network of 30 new satellites. This navigation system will be called Galileo.

This nonmilitary GPS receiver is showing a mapped route based on information gathered by GPS satellites.

It's Like Finding an Elephant in a Haystack

GPS doesn't just help people figure out where they are. It's a technology used to track animals as they migrate and move across Earth. Scientists attach small transmitters to wolves, bears, lions, and birds. Marine scientists have attached similar devices to seals, penguins, whales, and sea turtles. The transmitters send out a signal that is picked up by a GPS receiver. It is then coordinated with GPS satellites. This amazing technology makes it possible to follow animals on the go—even those that travel where people cannot.

This elephant has a GPS transmitter around its neck. The transmitter sends out a signal to GPS satellites, helping scientists keep track of the animal in the wild.

Earth Science Satellites

Remember *Landsat 1*? It was the first satellite designed to capture images of Earth's land areas back in the early 1970s. Well, *Landsat* is still going strong. In fact, it's the longest-running satellite program of its kind. *Landsat* satellites have acquired millions of images of our planet over a period of more than 30 years.

Landsat 7 is currently orbiting Earth. It produces a complete set of images of Earth's landmasses every 16 days. Each *Landsat 7* image covers more than 12,000 square miles (31,000 sq. km) of Earth's surface.

Scientists use *Landsat 7* data for making maps, tracking forest fires, finding minerals, measuring water quality in lakes, and protecting wildlife. *Landsat 7* images can also show farmers where and when to irrigate (water) their crops or which grazing land is best for their animals.

Since 1986, France's SPOT satellites have provided images of Earth's surface that rival those of *Landsat 7*. In fact, SPOT images show more detail.

The European Space Agency has launched its own Earth-observing satellites. So have Japan, Russia, India, and several other nations.

A U.S. military satellite took this image of Kabul, Afghanistan, following a U.S. bombing of the city in 2001. Military satellites also helped guide the attack.

The Big Picture

Some of the world's newest satellites are designed to study Earth as a global system. These satellites are collecting data on the atmosphere, the land, and the oceans. They also collect information about how these different parts of our planet interact with one another. Scientists use this information to study problems such as global warming and climate change.

The Sky Has "Eyes"

Not all the satellites orbiting Earth are designed for science. A number of countries have military satellites in space too. Many military satellites are similar to scientific or commercial satellites. But the information they send to Earth is in a code that only a special receiver can figure out. This is done to keep top-secret information from falling into an enemy's hands. Military surveillance satellites take more detailed pictures than scientific satellites do. They clearly "see" things about one foot in size—or even smaller. Such details can be important for armies tracking troop and tank movements or looking for land mines.

An EOS satellite took this image of the December 2004 tsunami (tidal waves) that struck parts of Asia and Indonesia. The image shows powerful deep ocean waves (*left*) off the coast of the island of Sumatra.

A satellite called *Terra* was the first of several satellites that are part of NASA's Earth Observing System (EOS) program. Launched in 2000, *Terra* is collecting data about Earth's land features as well as how much heat is absorbed and given off by the planet. *Terra* also gathers information about clouds, water vapor, snow, plants, soils, airborne particles, and gases such as carbon dioxide. It collects information about everything scientists think might help them better understand the big picture of how our planet works.

After *Terra* came *Aqua*. Also a part of NASA's EOS program, *Aqua* carries six instruments that observe the world's oceans, atmosphere, ice and snow, and its land and vegetation. *Aqua* takes its name from the Latin word for "water." It's a good choice, since one of *Aqua*'s biggest jobs is to gather information about Earth's water cycle.

Aqua also carries instruments that can determine air temperatures around the world—accurate to within one degree. It's a bonus that meteorologists hope they can use to improve weather forecasting even more.

And the List Goes On

Dozens of other satellites are up there sharing space with GOES satellites, *Landsat, Terra,* and *Aqua.* Each has a specific job. Here's just a sampling of some of the other satellites you'd encounter on a tour of the world at a couple hundred miles (about 300 km) above Earth's surface.

In 2004, *Aura* joined *Terra* and *Aqua* as the third satellite in the EOS program. *Aura* is a single-minded satellite. Its job is to measure certain types of gases in the atmosphere. Scientists hope this information will give them a better idea of chemical changes that are taking place in the air around us.

FUN FACT!

The awesome images taken by *Landsat 7* and other Earth-orbiting satellites are made of thousands of tiny picture elements, or pixels. The pixels are so small and so close together that they look connected. In a *Landsat 7* image, each pixel equals a piece of Earth's surface that is about 300 square feet (30 sq. m).

The key to *ICESat*'s job is in its name. This satellite is gathering information on ice—glaciers and ice sheets on land and sea ice on the oceans. Scientists hope it can help them figure out the role that polar regions play in Earth's climate.

Topex/Poseidon and *Jason* are ocean-ography satellites. They track ocean currents, changes in ocean level, and ocean winds. These are all things scientists need to figure out how the oceans affect the world's weather and climate.

GRACE is short for the Gravity Recovery and Climate Experiment. GRACE involves two satellites flying about 130 miles (209 km) apart in a polar orbit around Earth. Their mission is to sense slight differences in our planet's gravity. These differences help scientists better understand Earth's structure. They are also clues to certain kinds of changes taking place in the planet, such as how molten rock and other materials are moving around inside Earth.

An artist created this image of the two-satellite GRACE system (*above*), which took this image of Earth's gravity field (*right*).

QuickSCAT is a satellite that studies wind. It records how fast and in what directions winds are blowing across the surfaces of the world's oceans. Scientists use this information for weather forecasts, storm warnings, and global climate research. True to its name, *QuickSCAT* is a speedy satellite. It completes 14 orbits of Earth every day.

chapter 4

Space Science Satellites

So far you've learned about satellites that have their instruments focused on Earth. But there are many other satellites that are looking out into space. These space science satellites are exploring where no human being has yet gone.

SOHO is studying the Sun. The name stands for Solar and Heliospheric Observatory. Launched in 1995, *SOHO* is a joint effort between NASA and the European Space Agency. The satellite is in orbit between Earth and the Sun. It is about 950,000 miles (1.5 million km) from Earth, and 93 million miles (150 million km) from the Sun. Its mission is to unlock some of the Sun's many secrets.

The Sun produces all of the heat and light that we receive on our planet. It's not solid, like Earth, but a giant ball of gases. Nuclear reactions taking place within the Sun and at its surface are the source of its radiant energy.

SOHO is studying solar flares, the solar wind, and sunspots. Solar flares are huge eruptions of material that blast out into space from the Sun's surface. The solar wind is a flow of gases that streams off of the Sun in all directions with incredible speed—about 1 million miles per hour (1.6 million km/hr). Sunspots are mysterious dark spots on the Sun's surface that have powerful magnetic fields.

Why care about these Sun events? The solar wind, flares, and sunspots generate powerful radiation that can damage satellites and endanger astronauts. We even feel the effects of these events down on Earth. Solar flares and sunspots interfere with communications, navigation systems, and even the power grids that supply us with electricity.

Burn, Baby, Burn: *SOHO* and the Solar Inferno

The Sun's outer surface is constantly moving, like the top of a pot of boiling water but on a much, much larger scale. These movements actually produce sounds. Because of the vacuum of space, though, these sounds are trapped inside the Sun. (Even if they could reach Earth, the sounds are so low that we couldn't hear them.) The

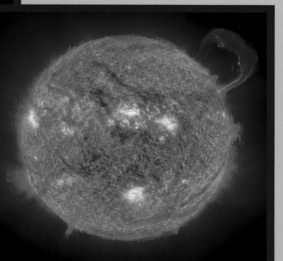

Solar and Heliospheric Observatory (SOHO), on the other hand, has instruments that can "hear" solar sounds—about 10 million separate notes. Scientists use information *SOHO* gathers about Sun sounds to learn more about the Sun's interior makeup and temperature.

SOHO photographed this image of the Sun during a solar flare (top, right).

Next Stop, Saturn!

A tiny satellite named *Cassini* is helping to answer one of the great questions about our solar system: why does the planet Saturn have such spectacular rings?

Launched in 1997, *Cassini* finally reached Saturn during the summer of 2004. It is orbiting Saturn, a planet that's only a bit smaller than Jupiter and has 31 known moons and those fantastic rings. Scientists know the rings are made of ice and rock. But there's so much they don't know. And they are counting on *Cassini* to provide a lot of answers.

Cassini captured this image of Saturn's rings (*below*) in 2004. The exploring orbiter also photographed Titan, Saturn's largest moon, the same year (*below, left*). Part of Titan's atmosphere can be seen at the top of the moon.

Cassini will spend four years orbiting Saturn. It will send back vast amounts of information about the planet and its rings. In 2005, an instrument called a probe was launched from the satellite. The probe landed on Titan, one of Saturn's moons.

Cassini is the result of an international collaboration between three space agencies. Seventeen nations helped build the satellite. More than 200 scientists worldwide will study the data that *Cassini* and its Titan-traveling probe send back to Earth.

A Space-Based Observatory

If you're interested in space at all, then you probably know something about the *Hubble Space Telescope.* About the size of a school bus, *Hubble* is a very special satellite. It's a space-based observatory, an orbiting telescope that's on a mission to explore the universe as never before.

In 1990, astronauts launched *Hubble* from the space shuttle *Discovery.* They put it in orbit roughly 380 miles (600 km) above Earth's surface. *Hubble* circles Earth every 97 minutes. From its spot in space, with no atmosphere to cloud the view, the *Hubble Space Telescope* has made some of the most amazing discoveries in the history of astronomy.

I Can See for Miles and Miles

The great advantage of having telescopes in space is there is no atmosphere to get in the way. When you look up at the night sky from Earth, you're looking through the atmosphere to see the stars. The atmosphere contains dust, water vapor, and other substances that can blur the view. That's why stars seem to twinkle when we look at them from the ground. The less atmosphere there is to look through, the better the image coming through a telescope. That's why many of the world's telescopes are on mountaintops or in such places as Antarctica, where the air is extremely dry and relatively pollution-free. But the best view of the stars is from outer space, far beyond Earth's atmosphere.

FUN FACT!

Hubble was the first satellite designed so it could be fixed and regularly updated by spacewalking astronauts. Space shuttle astronauts can move around next to *Hubble* and grab hold of it. They go out to make adjustments to the satellite and its instruments, and then return it to space.

Hubble can detect light with instruments that are five times sharper than the best ground-based tele-scopes. It can "see" more than visible light. *Hubble* can also see infrared and ultraviolet light, which are invisible to our eyes. High above Earth, *Hubble* has worked around the clock to provide stunning views of the universe that cannot be seen using ground-based telescopes or other satellites.

Astronauts repair the *Hubble Space Telescope* outside a space shuttle. It is very expensive to maintain *Hubble*, so its future is not clear.

Hubble's accomplishments are extraordinary. It has taken hundreds of thousands of images and observed tens of thousands of stars and other sites in the universe. And it has traveled more than 1.5 billion miles (2.4 billion km) since its launch.

Dazzling Discoveries

Hubble captured the best view of Mars ever obtained from Earth. It provided spectacular images of Comet Shoemaker-Levy 9 colliding with Jupiter. *Hubble* gave scientists their first detailed images of Pluto and its moon Charon and new information about the atmospheres of Uranus and Neptune.

The *Hubble Space Telescope* has helped astronomers understand how stars are born and eventually die. It has revealed how new planetary systems form. It traced mysterious gamma rays (radiation) to their origins in distant galaxies. *Hubble* provided proof that black holes exist and lie at

Hubble captured this image of the star V838 Monocerotis (center) in 2005. The star is 20,000 light-years from Earth.

the heart of most, if not all, galaxies. Data gathered by *Hubble* has helped scientists measure the age and size of the universe. The data has led many to believe that the universe is expanding faster and faster, driven by an unknown force.

In Good Company

Hubble is not alone in studying deep space. It is one of four satellite observatories that form NASA's Great Observatories Program.

The *Compton Gamma Ray Observatory* sensed gamma rays in space. It reentered Earth's atmosphere in 2000. The *Chandra X-Ray Observatory* detects X-rays. It can "see" much of the extremely hot matter in the universe that gives off X-rays but is otherwise invisible.

The 2003 *Spitzer Space Telescope* detects infrared light. It allows astronomers to view cooler objects in space, including those that are too dim to be detected by visible light alone.

Also orbiting out there is the *Far Ultra-violet Spectroscopic Explorer, or FUSE.* Scientists hope that this satellite will help them understand the origin and history of the chemical elements in the universe. They also hope it will show the processes involved in the evolution of galaxies, stars, and planetary systems.

FUN FACT!

Unlike the *Hubble*'s circular orbit that is relatively close to Earth, the *Chandra X-Ray Observatory* follows an elliptical orbit. Its distance from Earth changes over time. At its closest point, *Chandra* is about 6,000 miles (9,700 km) above Earth's surface. At its farthest point, it's 867,000 miles (140,000 km)—a third of the distance to the Moon.

Looking Ahead

Like satellites before them, those orbiting Earth or other planets will fulfill their missions and eventually fail. New satellites, with even better instruments, will be launched to replace them.

Already on the drawing board at NASA are plans for a satellite to replace *Hubble*. The James Webb Space Telescope is an orbiting observatory scheduled to head into space in 2011.

Also in the works are many new kinds of Earth-observing satellites. They will gather information about everything from how salty the oceans are to how much sunlight reflects off Earth's surface.

No one knows what the next generation of satellites will discover. But one thing is certain: the future of satellites looks as bright as the stars.

Glossary

altitude: the height of an object, such as a satellite or airplane, above the surface of a planet or another body in space

elliptical: shaped like an oval

geostationary: having a fixed orbit over the equator. *Geo-* refers to Earth, and *stationary* means "to have a fixed position." A geostationary satellite remains in place over the same spot on Earth.

Global Positioning System (GPS): a network of satellites used to pinpoint the location of something on Earth's surface

launch vehicle: a device, often a rocket, that carries a satellite into orbit

National Aeronautics and Space Administration (NASA): a part of the U.S. government that conducts space exploration and research missions. NASA was created in 1958, during the space race between the United States and the Soviet Union.

orbit: the path a satellite follows in space as it circles Earth or another object

polar orbit: an orbit in which a satellite travels around Earth from pole to pole

satellite: any smaller object that orbits a larger one

space race: beginning in the 1950s, a competition to develop rockets to send into outer space. The race was mainly between the United States and the Soviet Union, two powerful enemies at that time. The countries competed to be the first to send rockets to the Moon and to send human explorers into space. This competition resulted in many scientific and technical advances, including the invention of artificial satellites.

thrusters: small rockets that help steer and position satellites in space

Selected Bibliography

Harford, James J. "Korolev's Triple Play: Sputniks 1, 2, and 3." *NASA HQ History Office.* 2003.
http://www.hq.nasa.gov/office/pao/History/sputnik/harford.html (November 15, 2004).

"A History of Satellites and Robotic Space Missions." *Windows to the Universe.* 2002.
http://www.windows.ucar.edu/tour/link=/space_missions/unmanned_table.html (November 15, 2004).

NASA. "Communications Satellites." *NASA HQ History Office.* 2004.
http://www.hq.nasa.gov/office/pao/History/commsat.html (November 15, 2004).

NASA. "Facts & Figures." *HubbleSite.* 2005.
http://hubblesite.org/reference_desk/facts_.and._figures (November 15, 2004).

NASA. "40+ Years of Earth Science: 1958–1967." *NASA: Destination Earth.* 2003.
http://www.earth.nasa.gov/history/index10.html (November 15, 2004).

NASA. *NASA: Destination Earth.* 2004.
http://earth.nasa.gov/flash_top.html (November 15, 2004).

NASA. "The Solar and Heliospheric Observatory." 2005.
http://sohowww.nascom.nasa.gov/ (November 15, 2004).

NASA. "Spacelink—Satellites." *NASA Spacelink.* 2005.
http://spacelink.nasa.gov/Instructional.Materials/Curriculum.Support/Space.Science/Satellites/ (November 15, 2004).

National Geographic.com. "History of Satellites." *Eye in the Sky.* 2001.
http://www.nationalgeographic.com/eye/satellites.html (November 15, 2004).

"Navstar." *Encyclopedia Astronautica.* 2003.
http://www.astronautix.com/project/navstar.htm (November 15, 2004).

SARSAT. "NOAA SARSAT." *NOAA Satellites and Information.* 2005.
http://www.sarsat.noaa.gov/ (November 15, 2004).

Whalen, David J. "Communications Satellites: Making the Global Village Possible." *NASA HQ History Office.* 2005.
http://www.hq.nasa.gov/office/pao/History/satcomhistory.html (November 15, 2004).

Gaffney, Timothy R. *Secret Spy Satellites: America's Eyes in Space.* Springfield, NJ: Enslow Publishers, 2000. Gaffney's book provides a history of the United States' use of spy satellites, from the 1950s to current projects.

The Hubble Space Telescope Project. http://hubble.nasa.gov/. This NASA website explains the *Hubble*'s operations and technology. The site also includes the latest news of the project and an image gallery.

NASA: Earth Science. http://science.nasa.gov/EarthScience.htm. This NASA website provides information on earth science applications and technology, including satellites. It also offers activities, puzzles, and games designed for young readers.

NationalGeographic.com. "The History of Satellites." *National Geographic: Eye in the Sky.* http://www.nationalgeographic.com/eye/satellites.html. This site details the history of artificial satellites, from *Sputnik I* to the many satellites that currently circle Earth.

Parker, Steve. *Satellites.* Austin, TX: Raintree/Steck-Vaughn Publishers, 1997. This book discusses the history and development of artificial satellites.

The Solar and Heliospheric Observatory (SOHO) *Homepage.* http://sohowww.nascom.nasa.gov/. *SOHO* is a joint project between NASA and the European Space Agency. The website explains the mission and instruments used and provides a large gallery of images.

Walker, Niki. *Satellites and Space Probes.* New York: Crabtree Publishing Co., 1998. This book details several satellite missions and explains how satellite data is used in science and in everyday life.

Windows to the Universe: Chronology of Satellites and Robotic Space Missions. http://www.windows.ucar.edu/tour/link=/space_missions/unmanned_table.html. This website explains individual missions since the 1950s, from the United States, Russia, and other countries. The site also features images, news, and information on geology, astronomy, and physics.

Index

About the Author

Rebecca L. Johnson is a native of South Dakota, a state of prairie landscapes where night skies are exceptionally clear. She has written more than 50 books for children and young adults. Among her recent titles is a series about water biomes *(A Journey into the Ocean/a River/a Lake/an Estuary/a Wetland)* that received the Society of School Librarians International Best Book of 2004 award for science K-6. In addition to writing books, Johnson also works part-time as a writer for the U.S. Geological Survey's National Center for Earth Resources Observation and Science (EROS), a research center that houses more than 30 years of Landsat satellite images and continually downloads data from a variety of satellites orbiting the Earth.

Photo Acknowledgments

Photographs are used with the permission of: NOAO, background photos on pp. 1, 2, 4, 6, 8, 10-11, 12, 14, 16, 18, 20, 22, 24, 26, 28, 30, 32, 34, 36, 38, 40, 42, 44, 46, 48; NASA, pp. 5, 6 (top, middle and bottom), 7, 8, 12, 15, 16, 18, 21 (inset), 22, 23, 24 (top), 25, 26, 27, 28 (bottom), 34, 36 (main and inset), 38, 39 (left and right); © CORBIS, pp. 9 (left), 13, 24 (bottom), 28–29 (top), 41; © NASA/Roger Ressmeyer/CORBIS, pp. 9 (right), 19; © 1996 CORBIS/NASA, p. 10; Bill Hauser, pp. 11, 30; Library of Congress, p. 17 (LC-USZ62-10191); © Hulton-Deutsch Collection/CORBIS, p. 20; © Bettmann/CORBIS, p. 21 (main); © Getty Images, p. 29 (bottom); © DURAND PATRICK/CORBIS SYGMA, p. 31; © Reuters/CORBIS, p. 32; © CORBIS/SYGMA, p. 33; NASA, ESA, H. E. Bond (STScI), p. 42.

Front cover: © 1996 CORBIS/NASA (main); NASA (bottom left & right); © Anglo-Australian Observatory/David Malin Images (background). Back cover: © Anglo-Australian Observatory/David Malin Images (background).